Made in United States
Orlando, FL
20 April 2022

17022525R00061